Marcia da Silva Barros

The impact of land use on water quality

Marcia da Silva Barros

The impact of land use on water quality

Reflection of land use on the quality of surface water in the Ribeirão do Guaraçau catchment area, Guarulhos (SP)

ScienciaScripts

Imprint

Any brand names and product names mentioned in this book are subject to trademark, brand or patent protection and are trademarks or registered trademarks of their respective holders. The use of brand names, product names, common names, trade names, product descriptions etc. even without a particular marking in this work is in no way to be construed to mean that such names may be regarded as unrestricted in respect of trademark and brand protection legislation and could thus be used by anyone.

Cover image: www.ingimage.com

This book is a translation from the original published under ISBN 978-613-9-64563-3.

Publisher:
Sciencia Scripts
is a trademark of
Dodo Books Indian Ocean Ltd. and OmniScriptum S.R.L publishing group

120 High Road, East Finchley, London, N2 9ED, United Kingdom
Str. Armeneasca 28/1, office 1, Chisinau MD-2012, Republic of Moldova, Europe
Printed at: see last page
ISBN: 978-620-8-12181-5

ACKNOWLEDGEMENTS

I thank the great GOD for another stage of knowledge in my life.

I would like to thank Prof.^{0} Dr Antônio Roberto Saad, Prof.^{0} Dr Reinaldo Romero Vargas for their patience and dedication.

My special thanks go to Prof.^{a} Dr Regina Oliveira de Moraes Arruda who, when I needed it, kindly offered to help me, thank you very much.

I would like to thank my parents Judite and Zulmiro, and my beloved nephews Júnior, Gabriel, Natalia, Catarina, Danielle, Jonathan, Bernardo, Davi and the family and friends who have supported me on my journey. My safe harbour every day.

'Every second is time to change everything forever' (Charles Chaplin)

SUMMARY

The process of urbanisation that large urban centres have been undergoing has had a drastic impact on the quantity and especially the quality of water. The Guaraçau River Basin, located in the northern part of the municipality of Guarulhos, includes rural and urban areas with different land use classes. To assess environmental quality, physical-chemical analyses of temperature, pH, turbidity, conductivity and total phosphorus were carried out, as well as microbiological analysis (E. *coli*) over a 12-month period. The surface water of the Guaraçau stream in the rural area is already compromised, with a worsening in water quality from upstream to downstream for the parameters total phosphorus and E. *coli*, indicating faecal contamination due to the lack of basic sanitation in the region. Sites characteristic of rural areas are already showing serious signs of degradation, with trophic levels ranging from oligotrophic to hypereutrophic. This reinforces the need to build a sewage collection and treatment network at the Bonsucesso sewage treatment plant, which opened in 2011, and to control occupation in areas that produce good quality water.

Keywords: Urban waters. Eutrophication. Water pollution. São Paulo Metropolitan Region.

SUMMARY

CHAPTER 1

INTRODUCTION

The proximity between the municipality of Guarulhos and the capital of São Paulo acted as a favourable element for the expansion of the urban periphery of the municipality. The periphery is a location in constant movement, a space in continuous transformation. With the unbridled urban growth of the periphery of Guarulhos, which from 1950 onwards began to show increasingly high demographic numbers. It gained momentum with the organisation of allotments for the working class, who had no options in the urban centres. The allotments that sprang up on the outskirts of Guarulhos from 1950 onwards were completely lacking in infrastructure, with no water, sewage, paving, schools, electricity in the streets, and a lack of transport (MESQUITA, 2010).The unplanned growth of the urban outskirts of Guarulhos resulted in a process of environmental degradation, which affected most of the neighbourhoods and allotments located in its northern part. Graça (2007) subdivided the municipality of Guarulhos into two macro-compartments, separated from each other by the Jaguari River Fault, characterised by different lithologies, landforms and hydrography (Fig. 1 and Table 1).

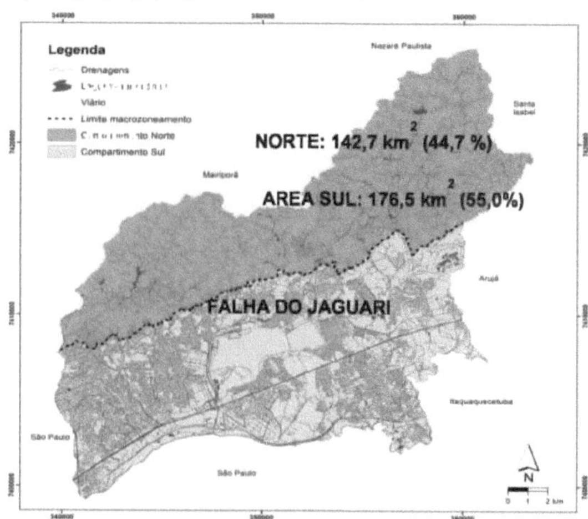

Figure 1 - Municipality of Guarulhos subdivided into two north-south macro-compartments Source: Graça (2007)

Table 1 - Main characteristics of the physical environment in the northern and southern macro-compartments of Guarulhos

Macro-compartments	Lithologies	Relief units	Drainage	Socio-economic
North	Precambrian metamorphic and igneous rocks	High and low mountains and hills	Dendritic pattern and high drainage densities	Developing rural areas

	Tertiary and Quartenary sedimentary rocks	Hills and river plains	Parallel pattern and low drainage densities	Urban and industrial areas
South	Tertiary and Quartenary sedimentary rocks	Hills and river plains	Parallel pattern and low drainage densities	Urban and industrial areas

Source: Lab. Geoprocessing Lab, 2015

The northern macro-compartment is dominated by relief units such as mountain ranges, high hills and low hills, while in the south there are hills, hills and river plains (ANDRADE; OLIVEIRA, 2008). In the south, there are flatter and more easily accessible areas, where the urbanisation process began. Today, this region is densely occupied with homes, industries on the banks of major motorways, an international airport, among other land use classes. The northern region, on the other hand, is made up of highly sloping terrain that is unsuitable from a geotechnical point of view, and is witnessing disorderly urbanisation with formal and informal allotments, often clandestine (ANDRADE PINTO et al., 2004).

The areas located in the northern macro compartment of the municipality lack basic infrastructure and present problems related to accentuated erosion processes, which, combined with unplanned urban growth, generate a high level of degradation of the physical and biotic environments.

1.1 The concept of water quality

When we use the term "water quality", we need to understand that this term does not necessarily refer to a state of purity, but simply to chemical, physical and biological characteristics, and that,

According to these characteristics, different purposes are stipulated for water. Thus, the national regulatory policy for water use, as set out in Resolution 20 of the National Environment Council (CONAMA), has sought to establish parameters that define acceptable limits.

The bodies of water were classified into nine categories, five of which were freshwater (salinity <0.5%), two saline (salinity greater than 30%) and two brackish (salinity between 0.5 and 30%). The "special" class is suitable for domestic use without prior treatment, while class IV domestic use is restricted, even after treatment, due to the presence of substances that pose a risk to human health. The standardised classification of water bodies makes it possible to set targets for achieving levels of indicators consistent with the desired classification (BRASIL, 1997).

Human and uninhabited occupation affects water quality due to unplanned urbanisation, the result of public policies in relation to the proliferation of clandestine allotments. *In natura* discharge of domestic and industrial effluents and agricultural activities. Disorganised exploitation of natural resources and inadequate land use (ANDRADE PINTO et al., 2004).

The Guaraçau River Basin is located in the northern macro-compartment of the municipality of Guarulhos and was an excellent setting for analysing the quality of its waters in relation to land use and occupation.

CHAPTER 2

OBJECTIVES

The aim of this research is to assess the stage of eutrophication of surface waters along the Ribeirão Guaraçau Hydrographic Basin (BHRG) in relation to the predominant land use conditions in the different compartments referring to the rural and urban areas.

2.1 Specific objectives

The specific objectives are:

a) To assess the behaviour of water quality in the BHRG through physical-chemical and microbiological parameters over a 12-month period, both in rural and urban areas;

b) Interpret the values analysed in the light of CONAMA legislation no. 357/2005 and use the classification according to State Decree 10.755/77;

c) Correlate the Trophic State Index (TSI) as a function of land use along the BHRG.

CHAPTER 3

THEORETICAL BACKGROUND

3.1 River basin as a unit of analysis

Historically, the river basin has been adopted as a preferential territorial unit for the study, planning, management and administration of water resources, as it allows for a perfect interaction between the characteristics of its physical and biotic environments with the various forms of land use, as well as natural resources, of which water stands out (MACHADO; TORRES, 2012).

The growing demand for all water uses (fishing, irrigation, electricity generation, public and industrial supply, and leisure) has made it possible in recent decades to draw up specific policies and legislation, while at the same time enshrining the river basin as a planning unit. This is the case, for example, with Federal Law No. 9.433 of 8 January 1997 (the Water Law), which established the river basin as the territorial unit for implementing the National Water Resources Policy (PNRH) and the National Water Resources Management System (SNGRH) (MACHADO; TORRES, 2012). In this way, we can see that the concept of the river basin is being broadened and consolidated as the preferred unit of study.

In recent decades, Brazil has experienced significant socio-economic development, with significant population growth which has led to the formation of metropolitan regions. In these megalopolises, urban and industrial expansion has led to an increase in demand for water resources. As a consequence, different types of environmental impacts have arisen, such as pollution resulting from the discharge of effluents (industrial and/or domestic) *in natura* into water bodies (TUCCI, 2008).

Mota (1988) draws attention to the fact that in water resource management proposals, special importance should be given to land use, with the aim of preventing the deterioration of a body of water's environmental health beforehand. The integration of water resource management with the environmental management of a river basin becomes a relevant fact, as analysing the changes in its natural landscape reveals an intimate relationship between them and the historical process of land use and occupation (TUCCI, 2005).

A hydrographic basin, also known as a drainage basin, corresponds to a portion of the

Earth's surface drained by a main river, its tributaries and sub-tributaries. The topography of the land is responsible for the drainage of water from pluviometric precipitation (rain) into this watercourse. The boundaries between river basins are found in the highest parts of the relief and are called watershed boundaries, as they separate the waters of different basins in which the surface runoff at any point converges towards a single fixed point, the outflow Fig. 2.

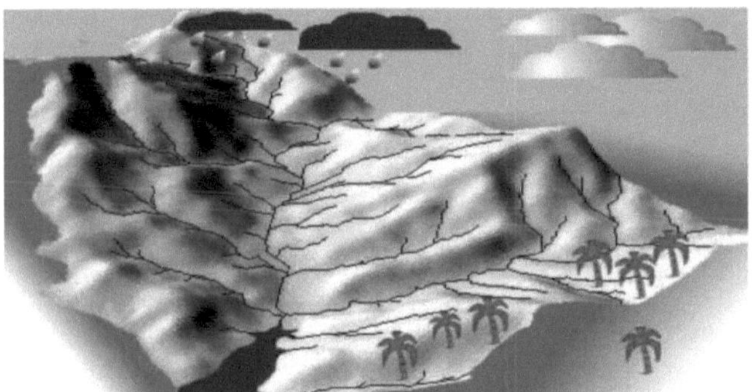

Figure 2 - Surface runoff at any point converges to a single fixed point, the outflow Source: Department of the Environment and Water Resources (2015)

The definition of a hydrographic basin is a set of areas with slopes in the direction of a given section where a watercourse is present, i.e. an area defined and topographically closed off from a watercourse (GARCEZ; ALVAREZ, 2002). Botelho (1999), on the other hand, simplifies this definition as being an area on the earth's surface that is drained by a main river and its tributaries, with its boundary being drawn by the so-called water dividers.

The advantages of working with river basins for interdisciplinary studies, management and planning are as follows:

a) the river basin is a physical unit with delimited boundaries;
b) has an integrated hydrological ecosystem, providing adequate management of water resources;

c) is a way of collecting and creating a database on biogeochemical, economic and social factors;
d) offers the opportunity for partnerships, problem-solving and, above all, encourages the participation of the local population in decision-making;
e) promotes concrete data for public decision-making, resulting in the management of the basin's sustainable development.

Garcez and Alvares (2002) point out that for integrated planning of the resources present in the basin, information must be collected on:

   a) quantity and quality of the water present;
   b) surveys of existing cartographic data;
   c) detailed information on the region's physical and biotic environment and socio-economic data.

As a result, the river basin has now become an ideal unit for studies, data collection and land use and occupation planning. It is also important to emphasise that, as it is a closed unit, it is possible to work in greater detail, known as "micro-watersheds".

The watershed must be carefully selected and present the physical and socio-economic conditions of the region. As an initial study, it is necessary to carry out a general preliminary survey of the local natural characteristics and then to analyse them in detail, as Garcez and Alvares (2002) have done:

   a) climate: reveals important information about the times of floods, frosts, droughts and the greatest potential for erosion processes;
   b) geology: the mineral, textural and structural characteristics of rock bodies, as well as the shape of the relief and soil, contribute to understanding the behaviour of exogenous processes;
   c) Relief: geomorphological information provides analysis of the landscape, showing factors of accumulation or transport of materials and anthropic actions;
   d) soils: shows the material to be eroded, transported and deposited in the short term, especially if there is unplanned anthropogenic action;
   e) drainage: reveals the availability of water resources;
   f) vegetation cover: they protect the soil from erosion and provide important information on land use and occupation;
   g) land use and occupation: types of occupation (formal and informal).

3.2 Water quality

According to Tucci (2008), the use of and competition for natural resources began in the middle of the 20th century, due to urban development and population growth. These developments have had an aggravating impact on much of the natural biodiversity of rivers, due to the pollution generated by human beings.

Water resources in Brazil's large urban centres are usually polluted due to the load of effluents (domestic and industrial) that they receive, as well as due to urban flooding, which mainly contaminates water sources. In addition to accelerated population growth and compromised river water quality, the reduction in vegetation cover is also a major factor leading to water scarcity. The sum of these factors then triggers a relative reduction in water availability, jeopardising regional development or, depending on the relevance of the water resource affected, even national development (BRAGA, 2003).

Physico-chemical and microbiological analyses of water can be used to assess the environmental quality of a river basin, as water acts as a geo-indicator, and once associated with other data, it allows environmental problems to be identified. Some of the physical-chemical and microbiological parameters analysed in water are shown in Table 2, where their respective environmental significance is summarised.

Table 2 - Analytical parameters and their environmental significance.

Turbidity (TU)
It can reduce the penetration of light, impairing photosynthesis due to suspended solids from the natural environment (rocks, clay, etc.) or man-made (domestic and industrial waste, etc.).
Temperature (T)
0 increase in temperature in water bodies can lead to an increase in chemical, physical and biological reactions and a decrease in dissolved gases such as DO, as well as the transfer of gases, causing a bad odour.
pH
Originates from the dissolution of rocks, oxidation of organic matter (o.m.) (natural) as well as domestic or industrial discharges (anthropogenic). Its rise may be associated with algae proliferation and its alterations are indicative of the presence of industrial effluents.
Total Phosphorus (PT)
Considered an essential nutrient for the growth of microorganisms responsible for stabilising organic matter. Usually found in sewage, associated with household chemicals (phosphate detergents) or of physiological origin (lipid phosphorus).
Dissolved Oxygen (DO)
Vital for aerobic aquatic beings, DO is the main parameter for characterising the effects of pollution in rivers by organic waste. High values are associated with the presence of algae and lower values are indicative of organic matter (sewage).
Total Solids (ST)
Total solids can be considered to be all the contaminants present in water, with the exception of dissolved gases. These solids can be volatile (organic) or fixed (inorganic). Their presence in rivers alters conductivity and pH values, as well as colour, which is associated with the nature of the contaminant.
Escherichia coli
Found in sewage, treated effluents and natural waters subject to recent contamination by humans and agricultural activities. This bacterium, which belongs to the faecal coliform group, poses health risks to humans and animals.
Conductivity
Conductivity is the numerical expression of water's ability to conduct electric current. It depends on ionic concentrations and temperature and indicates the amount of salts in the water column, thus representing an indirect measure of the concentration of pollutants.

Source: adapted from Sperling (2005).

3.2.1 Water pollution

Water quality can be affected by a wide range of human activities, whether domestic, commercial or industrial. With the growth in population and urban areas, controlling water resources has become complex over the course of human history. It is important to emphasise the existence of the following types of pollution sources (TUCCI, 1997): atmospheric, point, diffuse and mixed. Atmospheric pollution sources are classified as fixed (mainly industries) and mobile (motor vehicles, trains, aeroplanes, ships, etc.).

Point source pollution refers to pollutants that are discharged at specific points in water bodies in an individualised way, the emissions occur in a controlled manner and an average discharge pattern can be identified. Diffuse pollution occurs when pollutants reach bodies of water at random, and it is not possible to establish any discharge standards, whether in terms of quantity, frequency or composition. Rainwater run-off onto agricultural fields and accidents involving chemical or fuel products are examples of diffuse pollution. Mixed sources are those that include characteristics of each of the sources described by Tucci (1997): atmospheric, point and diffuse.

Studies on the quality of surface water in the municipality of Guarulhos, analysed in recent years (VARGAS et al., 2015; VARGAS et al., 2017) indicate that land use and occupation has compromised its quality, due to the consequences of the inadequate process of urban growth in the region, and that domestic and industrial pollution sources pose a considerable risk to water quality in the region.

3.2.1 Natural and Artificial Eutrophication

Eutrophication is the increase in the concentration of nutrients, especially phosphorus and nitrogen, in aquatic ecosystems, which has the effect of increasing their productivity (ESTEVES, 1988).

This process takes place mainly in lakes and reservoirs, although it can occur more rarely in rivers, as their environmental conditions are more unfavourable for algae growth.

There are several undesirable effects of eutrophication, including: bad odours and fish kills, changes in aquatic biodiversity, reduced navigation and transport capacity, changes in the quality and quantity of commercially valuable fish, contamination of water intended for public supply. The production of hydroelectric power can be affected by the excessive presence of aquatic macrophytes. In some cases, macrophytes can be present in the water after water treatment, which can aggravate their chronic effects.

3.2.2 Artificial Eutrophication

Esteves (1988) explains the process of eutrophication (eu = well; *trophos* = nutrients) as being the increase in nutrient concentration responsible for the increase in populations. In terms of the concentration of nutrients, mainly nitrogen and phosphorus, lentic ecosystems can be classified as: oligotrophic (low concentration), mesotrophic (medium concentration), eutrophic (high concentration) and hypereutrophic (very high concentration). In terms of its origin, the process can be classified as natural or artificial, i.e.

without or with anthropogenic action. When the process occurs without human intervention, it is slow, as the main agents are rainfall and surface runoff. When eutrophication arises from human action, it is called artificial, anthropic or cultural.

3.2.3 Trophic State Index (TSI)

The Trophic State Index aims to classify bodies of water into different degrees of trophicity, i.e. it assesses water quality in terms of nutrient enrichment and its effect on excessive algae growth or increased infestation of aquatic macrophytes. In this index, the results are obtained from phosphorus values and should be understood as a measure of the potential for eutrophication, since this nutrient acts as the causative agent of the process (LAMPARELLI, 2004).

In rivers, the EIT(PT) is calculated from the total phosphorus values using the formula according to Lamparelli (2004):

$$IET(PT) = 10.(6-((0.42-0.36.(ln.PT)/ln2)) \tag{1}$$

In lakes, the EIT (PT) is calculated from the total phosphorus values using the formula shown in equation 2, according to Lamparelli (2004).

$$IET(PT)= 10.(6-((0,42-0,36.(ln.PT)/ln2))-20 \tag{2}$$

Where total phosphorus (TP) is expressed in µg/L.

The EIT values are classified according to trophic status classes, shown in Table 3 below, along with their characteristics.

Table 3 - Trophic state class and its main characteristics.

EIT value	Trophic State Classes	Features
47	Ultraoligotrophic	Clean bodies of water with very low productivity and insignificant concentrations of nutrients that do not harm water use.
47<IET= 52	Oligotrophic	Clean, low-productivity bodies of water where there is no undesirable interference with water use due to the presence of nutrients.
52 <IET= 59	Mesotrophic	Bodies of water with intermediate productivity, with possible implications for water quality, but at acceptable levels in most cases.
59<IET=63	Eutrophic	Bodies of water with high productivity in relation to natural conditions, with reduced transparency, generally affected by anthropogenic activities, in which undesirable changes in water quality occur due to increased nutrient concentrations and interference with its multiple uses.

63<IET=67	Supereutrophic	Bodies of water with high productivity in relation to natural conditions, low transparency, generally affected by anthropogenic activities, in which undesirable changes in water quality often occur, such as algal blooms, and interference with its multiple uses.
>67	Hypereutrophic	Bodies of water significantly affected by high concentrations of organic matter and nutrients, with marked impairment of their uses, associated with episodes of algal blooms or fish kills, with undesirable consequences for their multiple uses, including livestock activities in riverside regions.

Sources: CETESB (2007) and Lamparelli (2004)

3.2.4 Legislation

According to Decree No. 10.755 of 22 November 1977 (BRASIL, 1977), which provides for the classification of receiving bodies of water in the classification provided for in Decree No. 8.468 of 08 September 1976, the following classifications apply to the study area: Baquirivu Guaçu River and all its tributaries, with the exception of the Tanque Grande Reservoir and its tributaries, up to the confluence with the Tietê River, in the municipality of Guarulhos - class 3. Table 4 shows the physical-chemical and microbiological parameters for waters classified as class 3, according to CONAMA Resolution 357/2005.

Table 4- Values of physico-chemical and microbiological parameters for class 3 waters according to CONAMA Resolution 357/2005

Physical, chemical and microbiological parameters	T°C	PH	PT (mg/L)	EC (µS/cm)	Escherichia coli (CFU/100 mL)	TU (UNT)
CONAMA 357/05 class 3	NE	6-9	Maximum 0.15 mg L^{-1}	NE	Maximum 2400 CFU 100mL^{-1}	Maximum 100 UNT

Source: Brazil (2005)

With regard to microbiological aspects, Brazilian legislation applies the concentration of faecal coliforms as the standard for the microbiological quality of surface water intended for public supply, recreation, irrigation and fish farming, and the concentrations of total coliforms and Escherichia coli as drinking water standards, according to CONAMA Resolution No. 357/2005 (BRASIL, 2005).

From the point of view of municipal zoning, the Ribeirão Guaraçau Hydrographic Basin (BHRG) is home to rural and urban areas that show different forms of land use, which can be sources of altered water quality in river courses. This basin was selected for investigation of the trophic state of its waters as a function of land use and occupation because it covers both rural and urban areas along its river course.

3.3 . Geoprocessing and land use mapping

The first studies on land use in Brazil began at the end of the 1930s and lasted until the 1940s, when studies on colonisation and reconnaissance trips predominated, such as those dedicated to analysing the colonisation of southern Brazil through migration or those that analysed the occupation of the Amazon (IBGE, 2006). The 1970s saw both advances in classificatory analyses of the forms and dynamics of land use, especially based on thematic focuses, and the use of statistical procedures in geography in technical and academic circles, reflecting a strong emphasis on quantitative analyses in the work produced at the time. The first systematic work using remote sensing as a tool for interpreting spatial phenomena of national significance was the Systematic Survey of Natural Resources, carried out by RADAMBRASIL, using radar images (SOBRINHO, 2013).

One of the most important points in Land Use is to gather information and create land cover patterns at the point studied. These actions involve office and field research, aimed at interpreting, analysing and recording observations of the landscape, concerning the types of land use and cover, with a view to classifying and spatialising them on maps. The survey of land use and cover includes analyses and mapping and is very useful for gaining up-to-date knowledge of the ways in which space is used and occupied, constituting an important planning and decision-making tool (SOBRINHO, 2013).

The different forms of use and occupation along river basins and other drainages can induce or even accelerate the process of erosion, both in the channel itself (smaller bed) and in the soil located in its beds (larger and exceptional), as in a significant area of a river basin. These activities are carried out for various purposes, such as in urbanised, rural, industrial and mixed areas. However, mixed areas were defined as having more impacting factors, as they have different activities in the same space, and a greater flow of people, especially when mixed in a rural use (permanent population) and in an industrial use (permanent population).

Although rural areas are less densely populated, economic activities in the countryside have a strong impact on the environment, especially when they suffer from deforestation problems, the use of heavy machinery, among others. However, when done with environmental planning and management, the damage is minimised (SOBRINHO, 2013).

By portraying the forms and dynamics of land occupation, these studies also represent a valuable tool for constructing environmental indicators and assessing the environmental support capacity, given the different management methods used in production, thus helping to identify alternatives that promote sustainable development (IBGE, 2006).

17

CHAPTER 4

CHARACTERISATION OF THE STUDY AREA

Guarulhos is one of the 39 municipalities of Greater São Paulo, Brazil's most economically important region. [a]It is the second most populous city in the state of São Paulo and the 12th most populous in the country (1,221,979 inhabitants) (IBGE 2006).

Thanks to various factors, such as the way it has been occupied, public policies and its location, Guarulhos has become a strategic distribution and logistics centre for the economy not only of the Alto Tietê region, but also of São Paulo and the country. The municipality is strategically located between two of the main national highways: Via Dutra, which links São Paulo to Rio de Janeiro, and Rodovia Fernão Dias, which links São Paulo to Belo Horizonte. It also has the Ayrton Senna motorway, one of the most modern in the country, which makes it easier to connect São Paulo directly with Guarulhos International Airport and the Port of Santos, from which it is 108 km away. The northern stretch of the Rodoanel is also under construction.

The Baquiruvu Guaçu River Basin (BHRG), which is the subject of this geoenvironmental analysis, covers an area of 149 km^2 and the Ribeirão Guaraçu River Basin is one of its sub-basins. The BHRG is inter-municipal and covers approximately 80% of the territory of Guarulhos. In the municipality of Arujá, it covers an area of 19.5 km^2 which corresponds to approximately 20% of its territory.

The studied area, represented by the Watershed, is located in the municipality of Guarulhos, covering the following neighbourhoods: Água Azul, Jardim Nova Bonsucesso, Jardim Ponte Alta, Mato das Cobras, Residencial Bambi and Vila Carmélia.

4.1 Location

The Ribeirão Guaraçau Hydrographic Basin (BHRG), Fig.3, lies entirely within the municipality of Guarulhos, which in turn is located in the northern sector of the São Paulo Metropolitan Region, approximately 17 km from the capital. It is bordered by the municipalities of Arujá (east), Itaquaquecetuba (southeast), Mairiporã (northwest), Nazaré Paulista (north), São Paulo (south and west) and Santa Isabel (northeast).

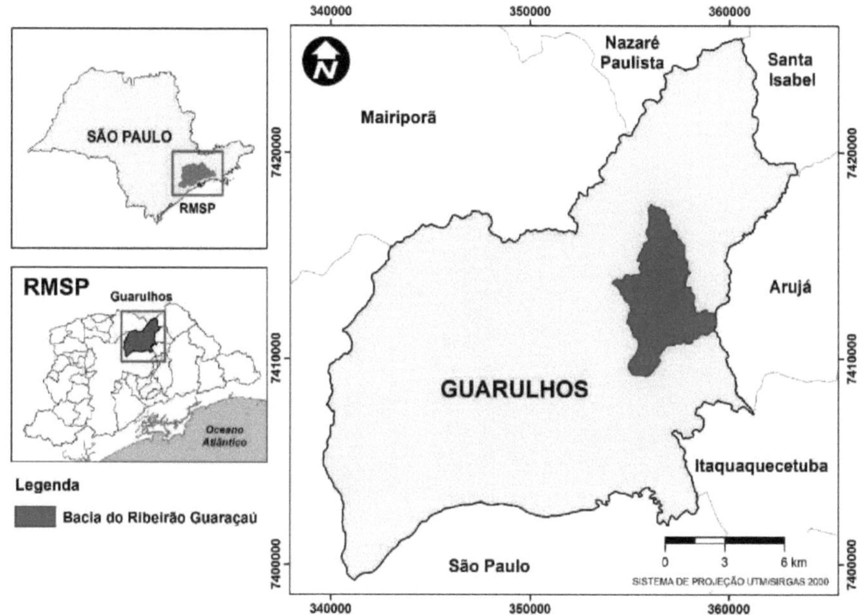

Figure 3- Location of the Ribeirão Guaraçau Hydrographic Basin (BHRG) in Guarulhos in the São Paulo Metropolitan Region (RMSP).Source: UnG Geoprocessing Laboratory (2015)

4.2 Characteristics of the BHRG

4.2.1 Hydrography of the BHRG

The municipality of Guarulhos is included in the Alto Tietê (UGRHI - 06) and Paraíba do Sul (UGRHI - 02) Water Resources Management Units.

Fig. 4 and Table 1 show the main river basins in the municipality.

The boundaries with the municipality of São Paulo are established, respectively, with the Cabuçu de Cima River to the west of Guarulhos and with the Tietê River to the south of Guarulhos. It also borders the municipality of Arujá via the Jaguari River. The Baquirivu-Guaçu River runs through the municipality.

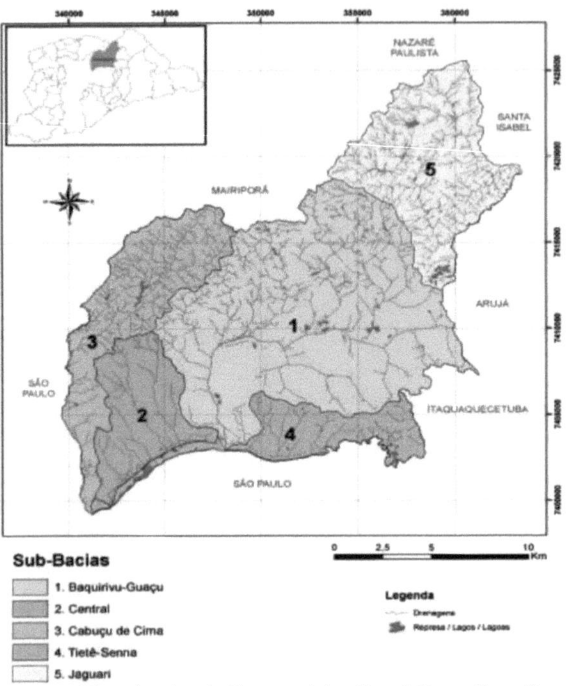

Figure 4- River basins in the municipality of Guarulhos Source: Andrade et al. (2008).

Table 1 - Main Basins in the Municipality of Guarulhos

BAY	SUB-BASIN	APPROXIMATE AREA	APPROXIMATE %
	Baquirivu-Guaçu	149 kilometres^2	46,6
	Cabuçu de Cima	49 kilometres^2	15,3
Upper Tietê	**Central**	**33 kilometres^2**	**10.3**
	Parati-mirim	4 kilometres^2	1.3
Paraíba do Sul	Jaguari	61 kilometres^2	19,0

Source: Andrade et. al. (2008)

The sub-basins in Guarulhos have different relief characteristics. The sub-basins of the Cabuçu de Cima river, the Jaguari stream and the contributing streams on the right bank of the Baquirivu-Guaçu river, have their headwaters formed in a relief of hills and mountains, with the channels embedded and strongly conditioned by tectonic structures, forming a dendritic to subparallel drainage pattern of high density. When they reach the region of the sedimentary basin where the relief is formed mainly by hills, like the other watercourses that emerge in this region, they change to a low drainage density, in wider valleys, sometimes with meandering channels on river plains (ANDRADE et al, 2008).

4.2.2 Guaraçau River Basin

The hydrography in the municipality of Guarulhos is divided into two basins: Alto Tietê and Paraiba do Sul:

a) Tietê basin with its sub-basins Baquirivu-Guaçu, Cabuçu de Cima, Central.

b) Paraiba do Sul Basin with its sub-basins Parati-mirim,Jaguari

The Ribeirão Guaraçau basin is located in the largest hydrographic basin in the municipality of Guarulhos, and occupies 46% of the 341 km2 of municipal territory.

Geometric features Area: 20.5 km^2 Length: 8350m , maximum width: 5600 m Maximum altitude: 990 m Minimum altitude: 750 m

4.3 Weather

Guarulhos lies in the transition between two of the planet's major climatic zones, delimited by the Tropic of Capricorn: the Tropical Zone and the Temperate Zone. The main characteristic of the region is the presence of a mild humid mesothermal climate, with one to two dry months. The average annual temperature is between 18°C and 19°C, with the coldest month averaging less than 15°C, while in the summer months the average varies between 23°C and 24°C. Rainfall is between 1,250 and 1,500 mm/year.

The basic characteristics are a dry winter and a rainy summer, influenced by Antarctic cold fronts and humidity from the Atlantic Ocean. The rainy season is marked by the occurrence of downpours that can reach accumulations of more than 50 mm in one hour, and flooding is common on these occasions. On the other hand, the air humidity can reach levels of less than 20 per cent, causing problems for public health and the risk of fires.

The most intensely urbanised area of the RMSP and Guarulhos is located in a sheltered region between the Cantareira Mountains to the north and the Mar Mountains to the south, where atmospheric circulation is not always favourable and there is a possibility of low air quality events, taking into account the huge amount of gases released by the metropolis' motor vehicles, industries and aircraft, further aggravated by the thermal inversion phenomenon that can occur in winter periods (ANDRADE et al., 2008).

CHAPTER 5

MATERIALS AND METHODS

5.1 Bibliographical research

The initial stage of the work involved bibliographical consultations, theoretical grounding and the assembly of a spatial database, carrying out a search for information on the area covered and the related themes in question. This information was based on dissertations, theses, books, technical reports, websites and scientific articles.

5.2 Land use map

Land use mapping was carried out in two stages, the first referring to photointerpretation and recognition of homogeneous land cover elements; the second corresponding to mapping by digitising the *layers* on the image in the spatial database.

In the photointerpretation stage, the visual aspects of the objects were identified and recognised based on the following parameters: colour, texture, geometry (shape), size, orientation and spatial distribution. Once the objects had been recognised, a hierarchical subdivision was adopted in the classification stage, based on the composition of the objects and their function in space.

Mapping from the *layers* on the digital database was carried out using *theArcGis software*. According to the classifications of Tominaga et al. (2004; 2005) and IBGE (2006), urban areas were analysed according to their stage of occupation (level of consolidation). Through the activity of photointerpretation and recognition of the objects observed in the land cover, land use mapping was carried out by classification. The land use map was drawn up with ten classes: tree cover, shrub vegetation, water bodies, farms, industrial warehouses, public facilities (schools, hospitals), consolidated residential areas, condominiums, unconsolidated residential areas and exposed soil.

5.   3Water Sampling   and Analysis

The samples were seasonal in order to describe the behaviour of the waters of the Guaraçau Stream between the dry and rainy periods, totalling 6 samples, between September 2015 and August 2016, through five (5) points chosen along the basin (Fig. 5).

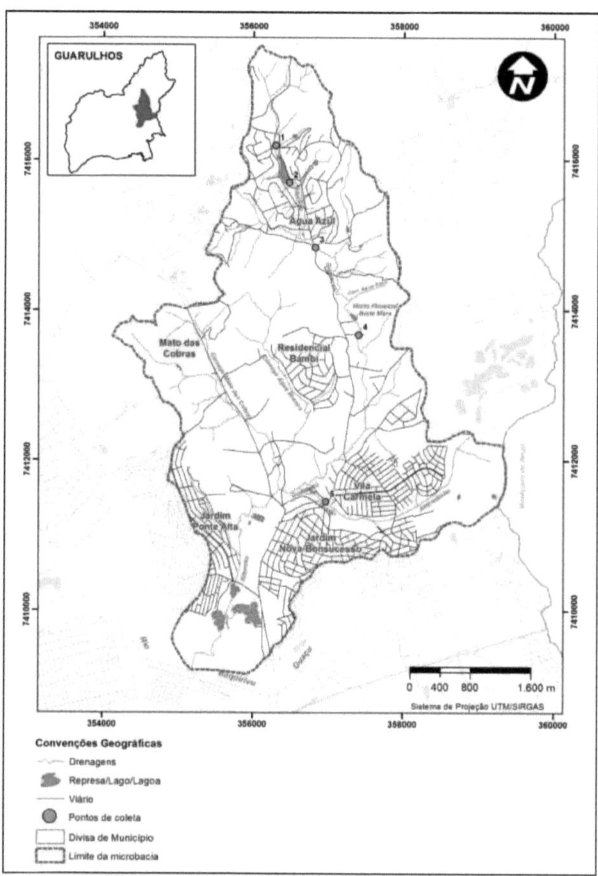

Figure 5- Water sampling points along the BHRG. Source: Lab. Geoprocessing Lab UnG, 2015

The sampling points and the corresponding Universal Transverse Mercator (UTM) are described in Table 2.

Table 2 - Universal Transverse Mercator UTM coordinates

POINT	LONGITUDE	LATITUDE
Point 1	23°21'26.93"S	46°24'22.60"O
Point 2	23°21'42.98"S	46°24'16.55"O
Point 3	23°22'14.45"S	46°24' 3.95"O
Point 4	23°22'50.74"S	46°23'44.67"O
Point 5	23°24' 3.14"S	46°24' 0.82"O

Source: from the author (2016)

The water was analysed in the field using duly calibrated analytical instruments and triplicate measurements for the following physical-chemical parameters: Temperature (using the Digimed DM-3 conductivity meter), pH (Digimed DM-2 pH meter), Conductivity (Digimed DM-3 conductivity meter), Turbidity (Quimis Q279P turbidity meter).

The other samples were collected in accordance with the National Guide for the Collection and Preservation of Samples (ANA, 2011) for the following chemical and microbiological parameters: Total Phosphorus (PT) and *Escherichia coli,* and were analysed according to the analysis methods described in Table 5.

The results of these analyses will be assessed against the standards established by CONAMA Resolution 357/2005 (BRASIL, 2005).

Table 5 - Chemical and microbiological variables and their respective analysis methods.
Source: Lab. Geoprocessing Lab, 2016

Parameters	Method of analysis
Total Phosphorus (PT)	APHA (2012). Standard Methods for the Examination of Water and Wastewater 22nd Ed - method 4500-P E
Thermotolerant coliforms (E. coli)	APHA (2012). Standard Methods for the Examination of Water and Wastewater (SMEVWV) 22nd Ed - method 9222

5.4 Calculation of the Trophic State Index (TSI)

The Trophic State Index aims to classify bodies of water into different degrees of trophicity, i.e. it assesses water quality in terms of nutrient enrichment and its effect on excessive algae growth or increased infestation of aquatic macrophytes. In this index, the results of the index calculated from the phosphorus values should be understood as a measure of the potential for eutrophication, since this nutrient acts as the causative agent of the process.

In rivers, the EIT(PT) is calculated from the total phosphorus values using the formula according to Lamparelli (2004):

$$EIT\,(PT) = 10.(6-((0.42-0.36.(ln.PT)/ln2)) \qquad \text{Equation (1)}$$

In lakes, the EIT (PT) is calculated from the total phosphorus values using the formula according to Lamparelli (2004):

$$IET(PT)= 10.(6-((0.42-0.36.(ln.PT)/ln2))-20 \qquad \text{Equation (2)}$$

Where total phosphorus (PT) is expressed in μg/L.

The EIT values are classified according to trophic status classes, as shown in Table 3, together with their environmental characteristics (LAMPARELLI, 2004).

CHAPTER 6

RESULTS AND DISCUSSION

The discussion and elaboration of the results of the proposals mentioned in the objective are presented in the article entitled **"Evaluation of water quality and trophic status in the Ribeirão Guaraçau catchment area, Guarulhos (SP): a comparative analysis between rural and urban areas".**

SUMMARY

The process of urbanisation that large urban centres have been undergoing has had a drastic impact on the quantity and especially the quality of water. The Guaraçau River Basin, located in the northern part of the municipality of Guarulhos, includes rural and urban areas with different land use classes. The aim of this study was to assess the water quality and diagnose the stage of eutrophication of the surface waters of Ribeirão Guaraçau, the main water body in the Ribeirão Guaraçau Basin. To assess environmental quality, physical and chemical analyses of temperature, pH, turbidity, conductivity and total phosphorus were carried out, as well as microbiological analysis (E. *coli*) over a 12-month period. The Trophic State Index (TSI) was used to ascertain the environmental degradation conditions of lotic and lentic environments. The surface waters of the Ribeirão Guaraçau in the rural area are already compromised, with a worsening in water quality from upstream to downstream for the parameters total phosphorus and *E. coli,* indicating faecal contamination due to the lack of basic sanitation in the region. Sites characteristic of rural areas are already showing signs of degradation, with trophic levels ranging from oligotrophic to hypereutrophic. This reinforces the need to build a sewage collection and treatment network at the Bonsucesso sewage treatment plant, which opened in 2011, and to control occupation in areas that produce good quality water.

Keywords: Urban waters. Water pollution. Eutrophication. São Paulo Metropolitan Region.

INTRODUCTION

The development of modern society, especially urban society, has taken place in a disorganised manner, without any planning, at the cost of increasing levels of environmental degradation. As a result of this unbalanced scenario, there are significant impacts that compromise environmental quality, especially in large metropolises. The municipality of Guarulhos, located in the Metropolitan Region of São Paulo (RMSP) and considered to be the second largest city in the state of São Paulo, is undergoing urban expansion and has planning problems induced by its industrial, road, airport and service development and significant civil works, such as the construction of the Rodoanel northern stretch (OLIVEIRA et al., 2009; MESQUITA, 2011).

The environmental impacts caused by the removal of forests, agricultural, urban and industrial development, combined with a lack of basic sanitation, leads to an increase in the concentration of nutrients in aquatic ecosystems, especially phosphorus and nitrogen, which results in the growth of algae, including potentially toxic cyanobacteria, posing a risk to the health of the ecosystem, as well as increasing the cost of water treatment for supply (ESTEVES, 2011). This process, known as eutrophication, can be quantified using the Trophic State Index (TSI), the purpose of which is to classify bodies of water into different degrees of trophy (LAMPARELLI, 2004). The index classifies the body of water into six trophic classes, according to the concentrations of total phosphorus in the water. The conditions conducive to the eutrophication process are those of a lentic environment, with the presence of nutrients, high temperatures, high levels of radiation, low turbidity and high water residence time. Water bodies with a lotic environment, classified as eutrophic, supereutrophic and hypereutrophic, rarely show eutrophication processes, but it is through rivers and streams that a large part of the nutrient input to lakes and reservoirs occurs (ANA, 2013).

Depending on land use and occupation, there are different levels of trophy in water bodies. Cunha et al. (2010)

studied three tropical rivers with different levels of anthropogenic interference. These authors observed that in more preserved regions, with 70% of their area occupied by forests, the trophic levels varied from oligotrophic to mesotrophic. On the other hand, in regions with the presence of industries, the trophic levels observed were hypereutrophic. In another study carried out in a river basin with a 65% urbanised area characterised by the presence of stilt houses, where domestic sewage is dumped *in natura* into the waters, the trophic levels ranged from supereutrophic to hypereutrophic. In the same study, in another river in the region with the presence of settlements and rural activities, trophic levels ranged from mesotrophic to supereutrophic (SILVA et al., 2014). Considering the concept of the hydrographic basin as a study unit (MACHADO and TORRES, 2012), the Ribeirão Guaraçau Hydrographic Basin (BHRG) was selected, located for the most part in the northern part of the municipality of Guarulhos, and which is a water-producing area in spring zones (ANDRADE et al., 2008). From the point of view of municipal zoning, there are rural and urban areas in this catchment, which show different forms of land use, which can be sources of alteration to the water quality of river courses. The main objective of this study was to assess the water quality and diagnose the stage of eutrophication of the surface waters of the Guaraçau Stream, the main body of water in the Guaraçau Stream Basin, in both rural and urban areas.

Geoenvironmental characteristics of the study area
The municipality of Guarulhos, located in the north-eastern part of the São Paulo Metropolitan Region (RMSP), is one of the 39 municipalities that make it up and is bordered by the municipalities of Arujá, Itaquaquecetuba, Mairiporã, Nazaré Paulista, São Paulo and Santa Isabel (Figure 1). The municipality is characterised by a humid sub-tropical climate with average annual rainfall of 1470mm. Annual rainfall in the region was 1897 mm in 2015 and 1570 mm in 2016 (INMET, 2016). Average annual temperatures in the colder months are between 17°C and 19°C, while in the summer, average annual temperatures vary between 23°C and 24°C (OLIVEIRA et al., 2009). In hydrographic terms, the municipality is divided into five basins, the largest of which is the Baquirivu-Guaçu Hydrographic Basin (BHBG) with an area of 149 Km^2 , in which the study area is located.
The geometric characteristics of the BHRG include an area of 20.5 km^2 , a length of 8350 m, a maximum width of 5600 m, a maximum altitude of 990 m and a minimum altitude of 750 m (RIBEIRO et a., 2013).

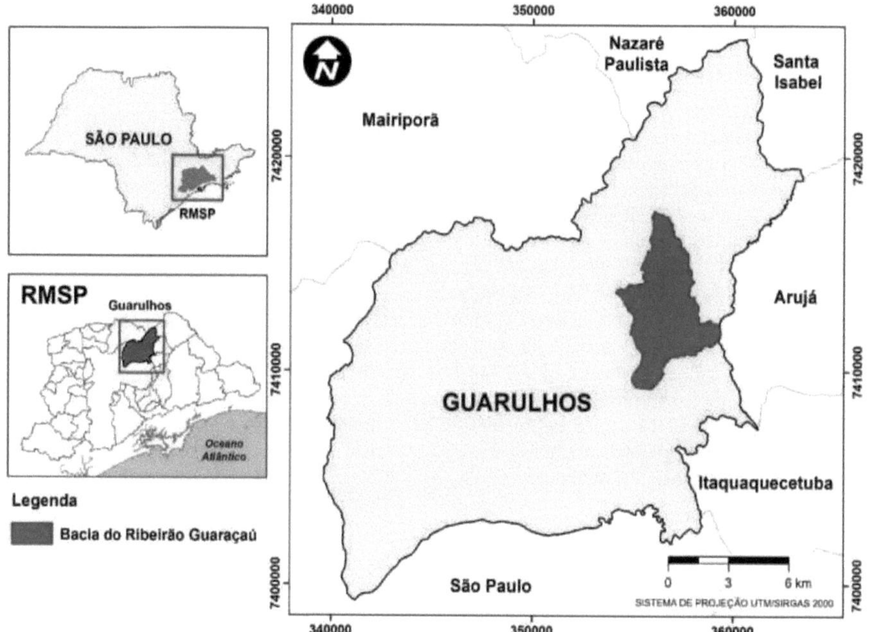

Figura 1. Location of the Ribeirão Guaraçau Hydrographic Basin (BHRG), Guarulhos (SP). Source: Lab. Geoprocessing Lab

The geometric characteristics of the BHRG include an area of 20.5 km^2 , a length of 8350 m, a maximum width of 5600 m, a maximum altitude of 990 m and a minimum altitude of 750 m (RIBEIRO et al., 2013). The landforms and lithological types present in the BHRG are summarised in Table 1.

Table 1. Characteristics of the physical environment of the Ribeirão Guaraçau Hydrographic Basin

ZONE	CHARACTERISTICS OF THE PHYSICAL ENVIRONMENT	
	Landforms	Lithological types
RURAL	Mountains and hills higher than lOOm	Predominance of metamorphic rocks (phyllites; ferriferous formation); igneous rocks (granites); localised alluvial sediments
	High hills with elevations above 900m	
	Morrotes	
	Restricted river plains	
URBAN	Low hills	Metasediments (phyllites).
	Morrotes	Elastic sediments (coarse to fine sandstones; claystones); alluvial sediments.
	Small hills	
	Wide and restricted river plains	

Source: Ribeiro et al., (2013)

2. MATERIALS AND METHODS

2.1. Land use map

The land use map was drawn up by updating the map from Ribeiro et al. (2013). The first step in this update was to export the vector file (shapefile) of this base map to *Keyhole Markup Language* (kml) format, in the ArcGIS programme - version 10 (ESRI, 2013). The next step was to import the kml file into *Google Earth,* where it was superimposed on an IKONOS II image, PSM Sensor 1 m resolution, from 28 September 2016, where the vertices of the vector were updated. In the final stage, the updated kml file was imported into ArcGIS, where it was exported to *shapefile* format.

2.2. Water collection and analysis

Five points (PI to P5) were selected to assess water quality along the Guaraçau stream, and six bimonthly samples were taken from September 2015 to August 2016. The selection of the location of the collection points (Figure 2) was based on the size of the drainage area and the occurrence of regions with different types of land use, with points PI to P4 located in an area with a predominance of rural areas and point P5 with urban characteristics. Point PI (23°21'26.93"S and 46°24'22.60"W) is located in a more preserved area, with a predominance of tree formation. Point P2 (23°21'42.98"S and 46°24'16.55"W) is located at the outlet of the Blue Lake. Point P3 (23°22'14.45"S and 46°24' 3.95"W) is located in the northern part of the catchment area and is influenced by several houses in the Agua Azul neighbourhood, where there is also a high-density urban slum with no sewage collection. Point P4 (23°22'50.74"S and 46°23'44.67"W) is located in an agricultural area, with a predominance of vegetated areas in its surroundings, but it already has contributions from the previous points. Point P5 (23°24' 3.14"S and 46°24' 0.82"W) is located in a high-density urban area.

The water was analysed in the field using duly calibrated analytical instruments and triplicate measurements for the following physico-chemical parameters:

Temperature (using a Digimed DM-3 conductivity meter), pH (Digimed DM-2 pH meter), Conductivity (Digimed DM-3 conductivity meter), Turbidity (Quimis Q279P turbidity meter). The other samples were collected in accordance with the National Guide for the Collection and Preservation of Samples (ANA, 2011) for the following parameters: Total Phosphorus and *Escherichia coli,* analysed in triplicate according to the *Standard Methods for the Examination of Water and Wastewater* (APHA, 2012).

The results of these analyses were assessed against the standards established by CONAMA Resolution 357/2005 (BRASIL, 2005) and in accordance with State Decree No. 10.755, which provides for the classification of receiving bodies of water under State Decree No. 8.468 of 8 September 1976, the Baquirivu Guaçu River and all its tributaries, up to the confluence with the Tietê River in the municipality of Guarulhos, were classified as class 3 (SÃO PAULO, 1977).

2.4. Trophic State Index (TSI)

For the waters of the Ribeirão Guaraçau, the EIT (PT) was calculated from the total phosphorus values (pg.L^{-1}) according to the methodology of Lamparelli (2004).

The EIT values and the six trophic status classes were distributed as follows (CETESB, 2007; LAMPARELLI, 2004): IET <47 (ultraoligotrophic); 47<IET < 52 (oligotrophic); 52 <IET < 59 (mesotrophic); 59<IET <63 (eutrophic); 63<IET <67 (supereutrophic) and IET > 67 (hypereutrophic).

3. RESULTS AND DISCUSSION

The Ribeirão Guaraçau Hydrographic Basin (BHRG) includes areas with both rural and urban characteristics. From point PI to point P4 there is a predominance of tree formation and undergrowth, low-density urban occupation, which characterises the region as a rural area (Figure 2). The consolidated urban area predominantly occupies the southern portion of the basin. Around point P5, the last collection point on the Ribeirão Guaraçau, there is the presence of orderly urban occupations with high density, essentially urban characteristics. Over the last ten years, the BHRG has undergone minor changes in terms of occupation (Figure 2) from 2007 to 2016. One negative point is that in this period, although high-density disorderly occupation (slums) contributed a small percentage, there was an increase of approximately 108% for this class, distributed throughout the BHRG, from its rural area to the urban area. The increase in these areas without sewage collection leads to a deterioration in the quality of the water bodies in the basin. Another relevant change for water quality was the 33 per cent reduction in agricultural cultivation areas. From the land use and occupation map, it can be seen that the majority of agricultural crops in 2007, reported by Ribeiro et al. (2013), have been replaced by lowland field vegetation, although this does not reflect an increase in the latter class, due to the 10% increase in low-density urban occupations in the rural region in these areas with lowland vegetation.

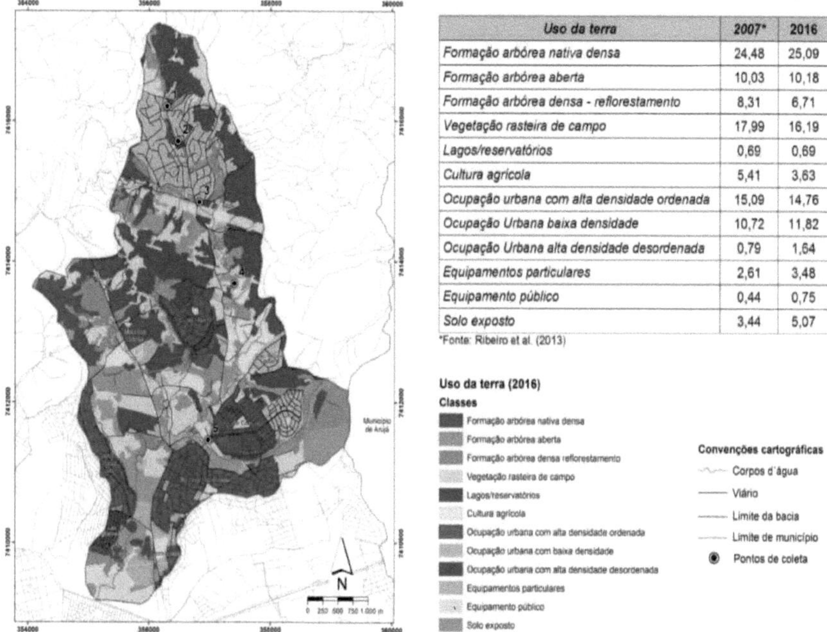

Uso da terra	2007*	2016
Formação arbórea nativa densa	24,48	25,09
Formação arbórea aberta	10,03	10,18
Formação arbórea densa - reflorestamento	8,31	6,71
Vegetação rasteira de campo	17,99	16,19
Lagos/reservatórios	0,69	0,69
Cultura agrícola	5,41	3,63
Ocupação urbana com alta densidade ordenada	15,09	14,76
Ocupação Urbana baixa densidade	10,72	11,82
Ocupação Urbana alta densidade desordenada	0,79	1,64
Equipamentos particulares	2,61	3,48
Equipamento público	0,44	0,75
Solo exposto	3,44	5,07

*Fonte: Ribeiro et al. (2013)

Figure 2.   Map of land use and occupation in 2016 with location of water sampling points and land use classes with their percentages (%) in the Ribeirão Guaraçau Hydrographic Basin (BHRG) in 2007 and 2016. Source: Lab. Geoprocessing Lab UnG, 2016

Point PI, located in a low-density urban area, has a more preserved area upstream with a predominance of trees. As a result, the values obtained in the surface waters analysed for all the parameters were of the highest quality (Table 1). Only the microbiological parameter (E.co/z) showed an average value above that established in the CONAMA resolution for class 3 water bodies (BRASIL, 2005). This high value is justified by the fact that there are farms with livestock around the point. The total phosphorus parameter for this point, despite being below the legal limit, had an average value close to the limit and a maximum value above the limit. This corroborates the faecal contamination of the water body, albeit to a lesser extent, probably caused by the faeces of farm animals. The behaviour of the water temperature (Table 1) varied according to the occupation of the surrounding area, with point PI having the lowest value because it refers to a more vegetated area, and point P5 having the highest average value because it is located in an urban area. Several studies indicate that the presence of a vegetated area around a water body causes water temperatures to be lower, while more urbanised areas with little vegetation increase water temperatures significantly (FIA et al., 2015; VARGAS et al., 2015; DE FREITAS PEREIRA et al., 2016; DE PAULA CARVALHO et al., 2016).

Table 1. Values of the physico-chemical and microbiological parameters of the waters along the BHRG from September 2015 to August 2016.

Parameters	Points	Average	dp	Min	Max.	CV (%)	CONAMA 357 Class 3
T(°C)	PI	18,5	4,2	13,0	24,3	22,9	N.E.
	P2	20,9	4,3	14,5	26,0	20,7	
	P3	20,7	4,0	14,9	24,9	19,5	
	P4	19,5	4,2	13,6	23,9	21,6	

Parameter	Point		SD			CV	Limit
	P5	21,8	4,0	15,3	25,6	18,5	
	PI	6,1	0,4	5,6	6,6	5,8	
	P2	6,9	0,4	6,5	7,5	5,8	
pH (upH)	P3	7,3	0,8	5,98	8,1	11,4	6a9
	P4	7,5	0,6	6,9	8,4	7,7	
	P5	7,0	0,7	6,5	8,3	10,7	
	PI	10	15	2	40	152	
	P2	22	20	2	55	88	
TU(UNT)	P3	30	11	12	39	36	max. 100
	P4	14	10	2	30	68	
	P5	**206**	204	32	532	99	
	PI	83	24	51	107	28,6	
EC (µS.cm)[1]	P2	138	23	115	172	16,5	
	P3	242	107	156	457	44,4	N.E.
	P4	860	415	446	1603	48,3	
	P5	671	235	421	956	35	
	PI	0,117	0,197	0,007	0,5	168	
	P2	**0,123**	0,285	0,007	0,71	231	0.05 (lentic)
PT (mg.L)[-1]	P3	**1,88**	2,1	0,007	5,14	112	
	P4	**2,54**	1,58	0,007	4,18	63	0.15 (lotic)
	P5	**4,84**	3,57	0,13	9,32	74	
	PI	**4,7E+03**	2,2E+03	2,5E+03	8,0E+03	47	
E. coli	P2	**6,3E+03**	4.1E+O3	2,4E+03	1,3E+04	65	
(CFU.100 ml	P3	**5,5E+05**	1.1E+O6	1,2E+04	2,8E+06	205	max. 2400
)[-1]	P4	**7,9E+05**	1.1E+O6	1,2E+05	3,0E+06	145	
	P5	**2,2E+07**	2,3E+07	3,0E+06	5,4E+07	104	

P2 lentic body (Blue Lake). Other points: lotic body
Abbreviations: SD (standard deviation); CV (coefficient of variation); T (temperature); TU (turbidity); EC (electrical conductivity); PT (total phosphorus); *E. coli {Eccherichia coli)*.

The pH analysis, for all the points analysed, showed no significant variation along the river basin, with the lowest coefficient of variation values, and their average values are within the limits set by CONAMA Resolution 357/05, even with areas with industrial and agricultural activity and a lack of basic sanitation in the region.

The turbidity values at points PI to P4 were below the maximum limit established by CONAMA 357/05, however, the turbidity at point P5 was above the established limit, due to the presence of domestic sewage and chemical industries in the region that discharge their effluents into the waters of the Guaraçau stream. The highest turbidity values were found during the rainy season, as also observed by Sardinha et al. (2008), Ríos-Villamizar et al., (2011), Fia et al. (2015) and Andrietti et al. (2016). During this period, suspended solids are swept away by surface runoff processes, especially in regions with exposed soils close to water bodies.

Numerous studies have reported the use of electrical conductivity in studies of the impact of pollutants on the aquatic environment, both in rivers (UWIDIA and UKULU, 2013; THOMPSON et al., 2012; VARGAS et al., 2015) and lakes (DAS et al., 2006; COSTA & HENRY, 2010). Each region has water with a characteristic electrical conductivity, depending mainly on the type of rock through which it permeates (TONG and CHEN, 2002). The conductivity values from points PI to P5 (Table 1) showed an increase in their average values, indicating an increase in the concentration of ions from upstream to downstream. One point worth highlighting is the high conductivity value (860 pS/cm) of point P4 located in an agricultural area. The use of fertilisers in the form of inorganic nitrogen and phosphorus salts causes the conductivity to increase significantly at this point. The contribution of other water bodies between point P4 and P5, through the process of dilution, causes the conductivity values to decrease at point P5 (671 pS.cm[-1]). Vargas et al. (2105) observed similar results in the increase of electrical conductivity values in the waters of the Taquara do Reino stream, in the catchment area of the same name and located in the north of the municipality of Guarulhos. According to the authors, the lack of basic sanitation in the region was the reason for the rise in electrical conductivity.

The waters of the Guaraçau stream join other contributors that flow into Lago Azul (Blue Lake), which originated as an old sand extraction pit (MESQUITA, 2011). This lake is currently used for recreational activities such as swimming and fishing by local residents. At this P2 point, the total phosphorus values for lentic bodies and, above all, the microbiological parameter were higher than those established by law. Faecal contamination from regular and irregular occupation, coupled with the lack of sewage collection and animal husbandry, leads to values above those permitted by legislation and certainly deserves greater attention from the municipal authorities, since the city council encourages the use of Lago Azul for recreational activities. Point P3, in the northern part of the catchment area, is located after the water bodies have been influenced by a number of homes in the Agua Azul neighbourhood, where there is even a high-density urban slum with no sewage collection. This scenario of water body degradation is confirmed when comparing point P2 with point P3 and observing an average increase of 15 times in the total phosphorus content and 87 times in the *E.coli* content. These values are continuously increasing throughout the basin, and the highest values were found at point P5, which is influenced by areas occupied by Chácara Recreio Rober and Vila Carmela.

The parameters total phosphorus and *E.coli* behaved differently when compared to the different rainfall periods. Total phosphorus showed higher values in dry periods, with the exception of point P4, indicating that in wetter periods there is a dilution effect of phosphorus in surface waters. The mean values and standard deviations of total phosphorus for the dry and rainy periods, respectively, were: PI (0.176 ± 0.281 and 0.059 ± 0.089), P2 (0.240 ± 0.403 and 0.007 ± 0.000), P3 (2.577 ± 2.335 and 1.175 ± 2.023), P4 (2.463 ± 1.056 and 2.607 ± 2.268) and P5 (5.350 ± 3.838 and 4.329 ± 4.035). The proximity of the phosphorus values at point P4 may be related to the constant use of fertilisers in the agricultural area (figure 2). Saad et al. (2013) observed this same dilution effect in the concentration of total phosphorus in rainier periods, and according to the authors this effect was more pronounced in consolidated urban areas with soil sealing. Rodríguez (2001) observed, as in this study, an increase in the concentration of nutrients in the water during the dry season and attributed this to the increase in their concentration in the water due to the reduction in flow. Similarly, Stacciarini (2002), assessing the quality of water resources in Paulínia (SP), observed that, at most of the points studied, phosphorus concentrations were higher during the dry season. Cruz (2003), assessing the waters of the Uberaba River (MG), found the same trends as this study in terms of total phosphorus levels in the rainy and dry periods. Farage et al. (2010) observed in the Rio da Pombas region (MG) that during the rainy season, the process of surface washing in anthropised areas provided a greater phosphorus load to the river, increasing its concentration, even with increases in the flow of the watercourse. The authors justified this increase in total phosphorus through the mechanism of surface runoff, which often occurs during rainy periods, especially in soils devoid of vegetation or with a predominance of ground cover, as is the case on the banks of the Pomba River. According to studies by Prada and Oliveira (2006), the increase in phosphorus in river water during the rainy season may also be related to the resuspension of bottom sediments as the flow rate increases. The quantity of microorganisms present in the surface water was higher in the rainy season, especially for the less preserved points, as can be seen in the logarithmic values of *E.coli* in the rainy and dry seasons, respectively, for P3 (5.36 ± 1.20 and 4.46 ± 0.14), P4 (5.96 ± 0.55 and 5.14 ± 0.10), and P5 (7.25 ± 0.41 and 6.92 ± 0.70). Saad et al. (2013) obtained similar results in the Ribeirão Tanque Grande watershed, which has similar geomorphological characteristics to the area studied, and due to the steeper slope there is greater surface water runoff carrying the microorganisms into the water bodies, compared to their infiltration into the soil (CHAVES and SANTOS, 2009; DE AZEVEDO LOPES and MAGALHÃES Jr., 2010; ANDRIETTI et al., 2016).

The Trophic State Index (Figure 3) was calculated from the total phosphorus values, with an increasing increase along the Ribeirão Guaraçau catchment area from the rural area to the urban area.

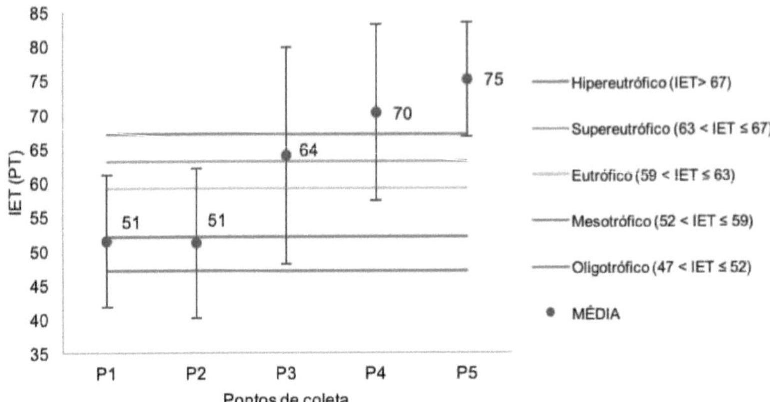

Figura 3. Behaviour of the EIT(PT) along the BHRG from September 2015 to August 2016

At the PI and P2 points, the best preserved, the level of trophic status varies from oligotrophic to mesotrophic, indicating the presence of clean bodies of water where there is no undesirable interference with water use due to the presence of nutrients, especially phosphorus. For the other points, it is possible to observe an increasing deterioration in the level of trophy of the water body (Figure 3). During the rainiest periods, the EIT (PT) showed more favourable trophic levels, i.e. lower, due to the nutrient dilution factor. For point P3, there was a predominance of the supereutrophic level, but with great variation from oligotrophic to hypereutrophic. The presence of housing estates and irregular occupation in the vicinity of point P3, associated with the temporal effect, as well as the presence of paving works in the area during the collection period, caused this great variation in the level of traffic. Similar results were observed by Saad et al. (2013) and Vargas et al. (2015) in watersheds in the municipality of Guarulhos. Point P4 showed a worsening in the level of trophy, moving to a hypereutrophic level, ranging from eutrophic to hypereutrophic. The presence of an agricultural area near this point leads to a worsening in the level of traffic due to the input of fertilisers used in the region's plantations (FARAGE et al., 2010; GONÇALVES e ROCHA, 2016; ZHOU et. al., 2016). Point P5 showed a hypereutrophic level for all determinations, indicating a body of water significantly affected by high concentrations of organic matter and nutrients, jeopardising the uses of this water.

4. FINAL CONSIDERATIONS

The recent crisis in the south-eastern region of Brazil from 2014 to 2016, which continues to affect much of the north-eastern region of Brazil, makes us reflect on the policy of preserving and monitoring river sources. The expansion of urban areas over these springs directly affects the quality of surface and groundwater, especially in developing countries, mainly due to the lack of sewage treatment, which is dumped directly into rivers, polluting and impacting the environment. The urbanisation process increases impermeable areas, leading to increased flooding and reduced infiltration into aquifers. The municipality of Guarulhos, located in the metropolitan region of São Paulo, has several water-producing areas in its northern portion, but unfortunately studies carried out in the municipality's river basins, especially in the Baquirivu Guaçu River Basin, show this scenario of water body degradation. In this study of the Guaraçau River Basin, the surface waters of the Guaraçau River can be seen to be compromised in the rural area around Lago Azul (P2). The other points, P3 to P5, showed a deterioration in water quality for the parameters total phosphorus and *E. coli,* indicating faecal contamination due to the lack of basic sanitation in the region studied. The increase in disorderly occupation in the area near Lago Azul contributed to the worsening of water quality. The values of both parameters for these collection points were above the limit established by CONAMA resolution 357/05 for class 3 water bodies. Points P3 and P4, which are characteristic of a rural area, showed signs of degradation with trophic levels ranging from oligotrophic to hypereutrophic for point P3, with a predominance of a supereutrophic level, and a worsening of point P4 with a predominance of a hypereutrophic level, ranging from eutrophic to hypereutrophic.

The Bonsucesso sewage treatment plant (ETE), located at the mouth of the Ribeirão Guaraçau catchment area, was inaugurated in December 2011 and is responsible for treating the sewage of 260,000 residents (SAAE, 2013). In this case, the need to build a collection network and sewage treatment at the Bonsucesso WWTP is reinforced (VARGAS et al., 2017).

5. ACKNOWLEDGEMENTS

To the Fundação de Amparo à Pesquisa do Estado de São Paulo (São Paulo State Research Foundation) for supporting the Research Grant Project, Process 2015/07406-4.

6. BIBLIOGRAPHICAL REFERENCES

ANA- NATIONAL WATER AGENCY. **Water resources in Brazil: 2013** Brasília: ANA, 2013.

ANDRADE, M.R.M; OLIVEIRA, A.O.; QUEIROZ, E.; SATO, S.E.; BARROS, E.J.; BAGATTINI, G.; ALEIXO A.A. Physiographic aspects of the Guarulhos landscape. *In* OMAR, E.E.H. (Org.) **Guarulhos has history:** questions about natural, social and cultural history. São Paulo: Ananda Gráfica e Editora, 2008. p. 25-37.

ANDRIETTI, G., FREIRE, R., DO AMARAL, A. G., DE ALMEIDA, F. T., BONGIOVANI, M. C., SCHNEIDER, R. M. indices of water quality and trophic state of Caiabi River, MT/Water quality index and eutrophication indices of Caiabi River, MT. **Revista Ambiente & Água,** v. 11, n. 1, p. 162 - 175, 2016.

APHA-AWWA-WPCF. **Standard Methods for Examination of Water and Wastewater.** 21 ed. Washington: [s.i.], 2012.

BRAZIL. CONAMA Resolution No. 357 of 17 March 2005. Provides for the classification of bodies of water and environmental guidelines for their classification, as well as establishing the conditions and standards for discharging effluents, and makes other provisions. **Federal Official Gazette.** Brasília, DF, 17 March 2005. Available at: <http://www.mma.gov.br/port/conama/res/res05/res35705.pdf>. Accessed on: 27 Apr. 2016.

CETESB. **Inland Water Quality Report for the State of São Paulo:** 2006. São Paulo: CETESB, 2007.

CHAVES, H. M. L.; SANTOS, L. B. Land use, landscape fragmentation and water quality in a small river basin. **Revista Brasileira de Engenharia Agrícola e Ambiental,** v. 13, p. 922-930, 2009. Available at <http://dx.d0i.0rg/1 0.1590/S1415-43662009000700015>

COSTA, M.L.R.; HENRY, R. Phosphorus, nitrogen, and carbon contents of macrophytes in lakes lateral to a tropical river (Paranapanema River, São Paulo, Brazil). **Acta Limnologica Brasiliensia,** v. 22, no. 2, p. 122-132, 2010.

CUNHA, D.G.F.; BOTTINO, F.; CALIJURI M.C., Land use influence on eutrophication- related water variables: case study of tropical rivers with different degrees of anthropogenic interference. **Acta Limnologica Brasiliensia,** v. 22, no. 1, p. 35-45, 2010.

CRUZ, L.B.S. **Environmental diagnosis of the Uberaba River basin, MG.** PhD thesis. Campinas: UNICAMP, 2003. 180 f.

DAS, R,; SAMAL, N.R., ROY, P.K.; MITRA, D. Role of Electrical Conductivity as an Indicator of Pollution in Shallow Lakes. **Asian Journal of Water, Environment and Pollution,** v. 3, n° 1, p. 143-146, 2006.

DE AZEVEDO LOPES, F. W.; MAGALHÃES JR., A. P. M. Influence of natural pH conditions on the water quality index (IQA) in the Ribeirão de Carrancas basin. **Geographies,** v. 6, n. 2, p. 134-147, 2010.

DE FREITAS PEREIRA, B. W., MACIEL, M. D. N. M., DE ASSIS OLIVEIRA, F., DA SILVA ALVES, M. A. M., RIBEIRO, A. M., FERREIRA, B. M., & RIBEIRO, E. G. P. Land use and water quality degradation in the Peixe-Boi River watershed, PA, Brazil/Land use and water quality degradation in the Peixe-Boi River watershed. **Revista Ambiente & Água,** v. 11, n. 2, p. 472, 2016.

DE PAULA CARVALHO, A., BALDUINO, Â. R., MACIEL, G. F., & PICANÇO, A. P. Assessment of pollution in rivers using water quality indices: a case study in Ribeirão São João in Porto Nacional-TO. **Geosciences (São Paulo),** v. 35, n. 3, p. 472-484, 2016.

ESTEVES, F.A. **Fundamentos de Limnologia.** 3ª ed. São Paulo: Interciência, 2011.

FARAGE, J. D. A. P., MATOS, A. T., SILVA, D. D., BORGES, A. C. Determination of the trophic state index for phosphorus in points of the Pomba River. **Revista Engenharia na Agricultura,** 18(4), 322-329, 2010.

FIA R.; TADEU H.C.; MENEZES, J.P.C; FIA, F.R..L; OLIVEIRA, L.F.C. Water quality of an urban lotic ecosystem. **Revista Brasileira de Recursos Hídricos,** v.20, n.2, p. 267-275, 2015.

GONÇALVES, D.R.P.; ROCHA, C.H. Water quality indicators and land use patterns in river basins in the State of Paraná. **Pesquisa Agropecuária. Brasileira,** Brasília, v.51, n.9, p.l 172-1183, 2016. DOI: 10.1590/S0100-204X2016000900017

INMET (National Institute of Meteorology). BDMEP - Meteorological Database   for Education and Research.   Available   at <http://www.ínmet.gov.br/portal/index.php?r=bdmep/bdmep>

LAMPARELLI , M. **C. Degree of trophy in water bodies in the state of São Paulo: evaluation of monitoring methods.2004. 235 f.** Thesis (Doctorate in Ecology) - University of São Paulo, São Paulo, 2004.

MACHADO, P. J. O.; TORRES, F. T. P. **Introduction to Hydrogeography.** São Paulo: Cengage Leaming, 2012.

MESQUITA, M. V. **Degradation of the physical environment in allotments in the Invernada, Fortaleza and Água Azul neighbourhoods, as case studies of urban expansion in the municipality of Guarulhos (SP).** PhD Thesis, Institute of Geosciences and Exact Sciences (Rio Claro), Julio de Mesquita Filho Paulista State University, UNESP, 2011.

OLIVEIRA, A. M. S.; ANDRADE, M. R. M.; QUEIROZ, W.; SATO, S. **Geoenvironmental Bases for an Environmental Information System for the Municipality of Guarulhos.** 2009. 179 p. Guarulhos: Guarulhos University. 2009 (FAPESP Report, process no. 05/57.965-1).

PRADA, S.M. ; OLIVEIRA, E. de Distribution of nutrients (C, N and P) in sediment samples from the Garças reservoir, Cotia - SP, In: 29ª Annual Meeting of the Brazilian Chemical Society, São Paulo, 2006. **Proceedings ...** Available at < http://sec.sbq.org.br/cd29ra/resumos/T1855-2.pdf>

RIBEIRO, T.F.B., ANDRADE, M. R.M., SATO, S. E., dos SANTOS, M. T.; SAAD, A. R. Geoenvironmental analysis of the Ribeirão Guaraçau catchment, Guarulhos (SP), based on the land use map and morphometric aspects, **Revista Universidade Guarulhos. Geociências,** São Paulo v.12 (1), p. 49 -62, 2013.

RÍOS-VILLAMIZAR, E. A.; MARTINS JÚNIOR, A. F.; WAICHMAN, A. V. Physical-chemical characterisation of waters and deforestation in the Purus River basin, Western Brazilian Amazon. **Revista Geografia Acadêmica,** v. 5, n. 2, 2011

SAAD, A.R. et al. trophic state index of the Ribeirão Tanque Grande catchment, Guarulhos (SP): comparative analysis between rural and urban areas. **Geosciences,** São Paulo, v. 32, n. 4, p. 611-624, 2013.

SÃO PAULO. Decree No. 10.755, of 22 November 1977. Provides for the classification of receiving bodies of water under Decree No. 8.468, of 8 September 1976, and makes related provisions. Official Gazette of the State of São Paulo, 1977.

SARDINHA, D. S.; CONCEIÇÃO, F. T.; SOUZA, A. D. G.; SILVEIRA, A.; JULI O, M. ; GONÇALVES, J. C. S. I. Evaluation of water quality and self-depuration of Ribeirão do Meio, Leme (SP). **Engenharia Sanitária e Ambiental,** v.13, p.329-338, 2008.

SILVA, G.S.; SANTOS E.A.; CORRÊA L.B., MARQUES A.L.B., MARQUES E.P., SOUSA E.R., SILVA, G.S. Integrated assessment of surface water quality: degree of trophy and protection of aquatic life in the Anil and Bacanga rivers, São Luís (MA). **Engenharia Sanitária Ambiental** v.19, n.3, p. 245-251, 2014.

STACCIARINI, R. **Evaluation of the quality of water resources in the Municipality of Paulínia, State of São Paulo, Brazil.** PhD thesis. Campinas: UNICAMP, 2002. 214f.

THOMPSON, M.Y.; BRANDES, D.; KNEY, A.D. Using electronic conductivity and hardness data for rapid assessment of stream water quality. **Journal of Environmental Management,** 104, p. 152-157, 2012.

TONG, S. T. Y.; CHEN, W. Modelling the relationship between land use and surface water quality. **Journal** of **Environmental Management,** New York, v. 66, p. 377-393, 2002.

UWIDIA, I. E; UKULU, H.S. Studies on electrical conductivity and total dissolved solids concentration in raw domestic wastewater obtained from an estate in Warri, Nigeria. **Greener Journal of Physical Sciences,** vol. 3 (3), pp. 110-114, April 2013.

VARGAS R. R.; SAAD A.R.; DALMAS F.B.; ROSA A.; ARRUDA R.O.M.; MESQUITA M.V.; ANDRADE M.R.M. Water Quality Assessment in the Córrego Taquara do Reino Hydrographic Basin, Guarulhos Municipality (São Paulo State - Brazil): Effects of Environmental Degradation. **Anuário do Instituto de Geociências** - UFRJ, Vol. 38 (2): 137-144, 2015.

VARGAS R.R., GONÇALVES J.J.S., DALMAS F.D., SAAD A.R., ARRUDA R.O.M., FERREIRA A.T.S. The contribution of the Guarulhos Municipality (São Paulo State) to the water quality of the Alto Tietê System. **Research in Geosciences,** 44 (1): 109-121 2017.

ZHOU P., HUANG J., PONTIUS JR R.G., HONG H., Newinsight into the correlations between land use and water quality in a Coastal watershed of China: Does point source pollution weaken it? **Science of the Total Environment** 543, 591-600, 2016.

CHAPTER 7

CONCLUSION

The Ribeirão Guaraçau Hydrographic Basin (BHRG) is part of the Baquirivu Guaçu River Basin in the municipality of Guarulhos and has both rural and urban characteristics. Over the course of a year, water samples were collected from five points in the BHRG, in an attempt to assess the effects of land use and the presence of vegetation cover on the water quality of the Guaraçau stream. From point PI to point P4, there was tree formation and undergrowth and low-density urban occupation, which characterises the region as a rural area, while around point P5, the presence of orderly, high-density urban occupation shows essentially urban characteristics.

In this study of the Guaraçau River Basin, the surface water of the Guaraçau River could be seen to be compromised in the rural area around Lago Azul (point P2). The other points, P3 to P5, showed a deterioration in water quality for the parameters total phosphorus and *E. coli,* indicating faecal contamination due to the lack of basic sanitation in the region studied. The increase in disorderly occupation in the area near Lago Azul contributed to the worsening of water quality. The values of both parameters for these collection points were above the limit established by CONAMA resolution 357/05 for class 3 water bodies. Points P3 and P4, characteristic of a rural area, showed signs of degradation with trophic levels ranging from oligotrophic to hypereutrophic for point P3, with a predominance of a supereutrophic level, and a worsening of point P4 with a predominance of a hypereutrophic level, ranging from eutrophic to hypereutrophic.

The Bonsucesso sewage treatment plant (ETE), located at the mouth of the Ribeirão Guaraçau catchment area, was inaugurated in December 2011 and is responsible for treating the sewage of 260,000 residents in the region (SAAE, 2013). In this case, the need to build a collection network and sewage treatment at the Bonsucesso ETE is reinforced.

REFERENCES

ANDRADE, M. R. M.; OLIVEIRA, A. M. S. Urban expansion and geoenvironmental problems of land use in Guarulhos. In: OMAR, E. E. H. **Guarulhos tem História.** São Paulo: Ananda Gráfica e Editora, 2008. p. 47-58.

ANDRADE PINTO, L. V. et al. Study of the springs in the Santa Cruz river basin, Lavras, MG. **Scientia Florestalis,** n. 65, 2004.

APHA-AWWA-WPCF. **Standard Methods for Examination of Water and Wastewater.** 21 ed. Washington: 2005.

BATTALHA, B-H. L.; PARLATORE, A. C. **Control of water quality for human consumption:** conceptual and operational bases. São Paulo: CETESB Library, 1977.

BOTELHO, R.G.M. Planejamento Ambiental em Microbacia Hidrográfica. In: GUERRA, A.J.; SILVA, A.S.; BOTELHO, R.G.M. (org.). **Erosion and soil conservation.** Rio de Janeiro: Bertrand Brasil, 1999. p.269-300.

BRAZIL. Decree No. 10.755 of 22 November 1977. Provides for the classification of receiving bodies of water under Decree No. 8.468, of 8 September 1976, and makes related provisions. **Federal Official Gazette.** Brasília, DF, 22 Nov. 1977. Disponível em <http://pnqa.ana.gov.br/Publicao/Decreto%20n%C2%BA%2010.755%20de%2022% 20de%20novembro%20de%201977.pdf>. Accessed on: 22 May 2016.

. CONAMA Resolution No. 357 of 17 March 2005. Provides for the classification of bodies of water and environmental guidelines for their classification, as well as establishing the conditions and standards for discharging effluents, and makes other provisions. **Federal Official Gazette.** Brasilia, DF, 17 March 2005. Available at: <http://www.mma.gov.br/port/conama/res/res05/res35705.pdf>. Accessed on: 27 Apr. 2016.

CETESB. **Inland Water Quality Report for the State of São Paulo:** 2006. São Paulo: CETESB, 2007.

ESTEVES, F.A. **Fundamentos de Limnologia.** Rio de Janeiro: Interciência, 1988.

GARCEZ, L.N.; ALVAREZ, G.A. **Hidrologia.** São Paulo: Edgard Blucher, 2002.

GRAÇA, B. A. **Geoenvironmental conditioning factors in the historical process of land occupation in the municipality of Guarulhos, State of São Paulo.** 2007. 147f. Dissertation (Master's in Geoenvironmental Analysis) - Guarulhos University, Guarulhos, 2007.

BRAZILIAN INSTITUTE OF GEOGRAPHY AND STATISTICS-IBGE. **Technical manual on land use.** Technical Manuals in Geosciences, number 7. IBGE, 2006.

LAMPARELLI , M. C. **Degree of trophy in water bodies in the state of São Paulo: evaluation of monitoring methods.** 2004. 235 f. Thesis (Doctorate in Ecology) - University of São Paulo, São Paulo, 2004.

MACHADO, P. J. O.; TORRES, F. T. P. **Introduction to Hydrogeography.** São Paulo: Cengage Learning, 2012.

MESQUITA, M. V. **Degradation of the Physical Environment in allotments in the Invernada, Fortaleza and Água Azul neighbourhoods, as case studies of urban expansion in the municipality of Guarulhos (SP).**2010. 145p. Thesis (Doctorate in Geosciences and Environment) - Institute of Geosciences and Exact Sciences, Paulista State University, Rio Claro, 2010.

MOTA, S. **Preservation of Water Resources.** Rio de Janeiro: ABES, 1988.

PIRES DO RIO, G. A.; GALVÃO, M. C. C. Gestão ambiental: apontamentos para uma reflexão. **Anuário do Instituto de Geociências,** Rio de Janeiro, n. 19, 1996.

PIRES DO RIO, G. A.; PEIXOTO, M. N. O.; MOURA, V. P. Water resources management: difficulties of territorial articulation. In: SIMPÓSIO DE RECURSOS HÍDRICOS DO CENTRO-OESTE,2., 2002. **Proceedings...** Campo Grande:[s.n.], 2002.

. Water Law: developments for environmental and territorial management. In: MATA, S F et al. **Educação ambiental:** projetivas do século. Rio de Janeiro: MZ Editora, 2001. p. 93-99

PORTO, A. A. **Land Use and Sewage Contamination of the Capão da Sombra Stream, Guarulhos, SP.** 2013. 85 p. Dissertation (Master's in Geoenvironmental Analysis) - Guarulhos University, Guarulhos, 2013.

RIBEIRO, T.F.B. et al. Geoenvironmental analysis of the Ribeirão Guaraçau catchment, Guarulhos (SP), based on the land use map and morphometric aspects. **Revista UnG - Geociências,** v.12, n.1, p. 49-62, 2013.

SAAD, A.R. et.al. Geoenvironmental conditioning factors in the historical process of territorial occupation of the municipality of Guarulhos, State of São Paulo, Brazil. **Revista Ung- Geociencias,** v.6, n.1, p. 12-23, 2007.

SPERLING, M. von., 2005. **Introduction to Water Quality and Sewage Treatment: principles of biological wastewater treatment.** Belo Horizonte: DESA/UFMG.

SOBRINHO, E. J. A. **Proposal for a geoenvironmental assessment system for rivers: case study for the Córrego das Cruzes (Santo Antonio do Arancanguá / SP).** São Carlos - SP, 2013. 230 f. Test (Doctorate) - São Carlos School of Engineering. University of São Paulo. São Carlos. 2013.

TUCCI, C. E. M. **Hidrologia:** ciência e aplicação. 2. ed. Porto alegre: ABRH/ editora da UFRGS, 1997.

. **Hidrologia:** ciência e aplicação. 2. ed. Porto alegre: ABRH/ editora da UFRGS, 2008.

. **National Rainwater Programme.** Brasilia: Ministry of Cities, 2005.

Printed by Books on Demand GmbH, Norderstedt / Germany